Cássio Ricardo Gonçalves da Costa
Stella R. P. Suddarth
Ailson de L. Marques

Land use: physical-chemical, biological and social aspects

Land use: physical-chemical, biological and social aspects

Cássio Ricardo Gonçalves da Costa
Stella R. P. Suddarth
Ailson de L. Marques

Management practices and use of soil and natural resources

Imprint
Any brand names and product names mentioned in this book are subject to trademark, brand or patent protection and are trademarks or registered trademarks of their respective holders. The use of brand names, product names, common names, trade names, product descriptions etc. even without a particular marking in this work is in no way to be construed to mean that such names may be regarded as unrestricted in respect of trademark and brand protection legislation and could thus be used by anyone.

Cover image: www.ingimage.com

This book is a translation from the original published under ISBN 978-613-9-75188-4.

Publisher:
Sciencia Scripts
is a trademark of
Dodo Books Indian Ocean Ltd. and OmniScriptum S.R.L publishing group

120 High Road, East Finchley, London, N2 9ED, United Kingdom
Str. Armeneasca 28/1, office 1, Chisinau MD-2012, Republic of Moldova, Europe
Printed at: see last page
ISBN: 978-620-6-43518-1

TABLE OF CONTENTS:

CHAPTER 1

SOIL MINERALOGY AND ITS EFFECT ON THE SOIL MICROBIAL COMMUNITY

Cassio Ricardo Gonçalves da Costa

Ailson de Lima Marques

Stella Ribeiro Prazeres Suddarth

INTRODUCTION

When studying mineralogy in the area of soils, one thinks directly of a study based on the ability to identify the origin and evolution of soils through the identification of minerals, verification of the ability to adsorb nutrients, through the origin and presence of charges, and the mineralogical composition of the soils that intends to use.

It is a fact that minerals are present in the composition of soils (Barré et al., 2014), but the lack of in-depth knowledge of their mineralogical composition highlights inaccurate and defective agriculture, which ranges from reducing microbial survival (Roco et al., 2017) to reducing the entire system's capacity to absorb nutrients (Rodrigues et al., 2016) in the various agroecosystems.

How can the mineralogical composition of the solid phase of soils influence the productive capacity and sustainability of agroecosystems?

Since an agro-ecosystem is an open system, interacting with nature, modifications are becoming more and more accentuated in order to strike a balance with society in each region. In order to modify an area into an agricultural system, recognising the area in relation to the history of land use is a starting point and most soils in tropical regions have a high degree of weathering, i.e. a higher concentration of 2:1 clay minerals, oxides and hydroxides. In these soils, organic matter is reduced and, since soil organic matter is a fundamental component in the functionality of the system, its reduction would influence the physical structure and environmental conditioning for heterotrophic organisms, as well as the carbon cycle of soils (Barré et al., 2014).

Mineral soils with clays of the 2:1 type are soils that have clay materials with particles smaller than 2pm and this strongly affects the dynamics of soil organic matter and consequently the microstructure and pore system of the soils. From this, it is possible to identify the carbon content and stability of the soil through the size of the clay particle, as observed by Feng et al. (2013).

Based on particle size, certain types of minerals can be observed: phyllosilicates (clay minerals), metal oxides and hydroxides (goethite, ferridite, etc.), primary minerals (quartz, feldspars, etc.) and in soils carbonates (gypsum).

The mineralogy of phyllosilicates (also known as clay minerals) is linked to the stabilisation of soil organic matter, as they are the main components of the clay fraction in most soils. Phyllosilicate soils can have three types of origins: they can have originated from parent materials, transformed from other minerals or neosynthesised. Phyllosilicates are divided into two types: those with 1:1 layers and those with 2:1 layers. Isomorphic substitution of cations is common in 2:1 clays and less common in 1:1 clays. Both 2:1 and 1:1 clays have pH-dependent charges, corresponding to the surface of the hydroxides located at the broken edges of the clay layers, which can be ionised depending on pH conditions (Barré et al., 2014).

Soils from tropical agroecosystems show a dominance in kaolinite (1:1), which may be more often associated with phyllosilicates of the 2:1 type (smectites, illite, chlorite), (Hong et al., 2012). In temperate soils, the clay fractions are generally dominated by 2:1 clay minerals and contain some kaolinite. From this point on, a greater absorption of organic matter in the soil can be defined, as phyllosilicates can absorb a variety of organic molecules in different mechanisms and this will determine the capacity of soils to supply nutrients to plants and organisms.

According to Bauer,Velde (2016), organic molecules can interact with two different regions of phyllosilicates: the basal planes and the edge sites. The reactivity of the basal planes will depend on the charge layer, and due to isomorphic substitution, almost all 2:1 clay soils are permanently negatively charged and can promote the adsorption of organic components on the phyllosilicate layers through various components.

At the broken edges of the clay layers, the surfaces of the hydroxides (Si-OH and Al-OH) can establish covalent bridges with the organic molecules and this ligand exchange interaction has been recognised as providing very efficient stability to organic matter from microbial decomposition (Mikutta et al., 2007). This stability of organic matter through phyllosilicates influences the better efficiency of liming and the greater availability of phosphorus in tropical agroecosystems that are deficient in this element.

Phosphorus in agricultural soils is an important limiting nutrient for crop production and due to its scarcity, especially in tropical agroecosystems, it is a major challenge for soil science in the 21st century. In addition to analysing soil fertility, mineralogy plays an important role in the amount of phosphorus in the soil, as the degree of interaction between phosphate and soil minerals can be measured via the maximum phosphorus absorption capacity or remaining phosphorus, with a positive correction between P sorption and the soil's clay fraction (Fink et al., 2016). This fact is interesting, especially when making decisions about fertiliser calculations and chemical conditioning of weathered soils, generating greater savings for the producer.

In research carried out by Fink et al. (2016), it was observed that the mineralogy of the soils under study had a direct influence on P adsorption capacity, which was positively correlated with the content of iron oxides, in particular with the amount of goethite, i.e. the greater the presence of goethite. On the other hand, remaining phosphorus was negatively correlated with goethite content, showing the direct effect of mineralogy on soil fertility.

The mineralogy of soils influences the reactivity of dust gases. Mineralogy, and in particular the amount and solubility of iron, is key to the impact of dust on marine biogeochemistry. According to Journet et al. (2014) the mineralogy of airborne dust is linked to the mineralogy of the fractions eroded from soil sources and particle size during emission. The most abundant minerals in soil and air are kaolinite, illite and smectite, which account for 68.6% of soil and 64.5% of air dust, mainly from arid and semi-arid regions. Knowledge of the mineralogical composition of the air helps to observe the agricultural impact of the region and to observe past and future climate change.

Soil microbiology develops under favourable temperature, humidity and nutritional conditions for each community. Soil mineralogy has a direct effect on the soil microbial community. The size of the particles, such as 2:1 clay minerals, can influence the pore system (Barré et al., 2014), i.e. greater microporosity and thus changes in the microbial community.

REFERENCES

Barré, P., O. Fernandez-Ugalde, I. Virto, B. Velde and C. Chenu (2014). "Impact of Phyllosilicate Mineralogy on Organic Carbon Stabilisation in Soils: Incomplete Knowledge and Exciting Prospects." **Geoderma**: 382-395.

Bauer, A. and B. D. Velde (2016). Geochemistry at the Earth's Surface, Springer Feng, W., A. F. Plante and J. Six (2013). "Improving Estimates of Maximal Organic Carbon Stabilisation by Fine Soil Particles." **Biogeochemistry:** (1-3): 81-93.

Fink, J. R., A. V. Inda, J. Bavaresco, V. Barrón, J. Torrent and C. Bayer (2016). "Adsorption and Desorption of Phosphorus in Subtropical Soils as Affected by Management System and Mineralogy." **Soil and Tillage Research** : 62-68.

Hong, H., G. J. Churchman, Y. Gu, K. Yin and C. Wang (2012). "Kaolinite-Smectite Mixed-Layer Clays in the Jiujiang Red Soils and Their Climate Significance." **Geoderma**: 75-83.

Journet, E., Y. Balkanski and S. P. Harrison (2014). "A New Data Set of Soil Mineralogy for Dust-Cycle Modelling." **Atmospheric Chemistry and Physics:** (8): 3801-3816.

Mikutta, R., C. Mikutta, K. Kalbitz, T. Scheel, K. Kaiser and R. Jahn (2007). "Biodegradation of Forest Floor Organic Matter Bound to Minerals Via Different Binding Mechanisms." Geochimica et **Cosmochimica Acta:** (10): 2569-2590.

Rodrigues, M., P. S. Pavinato, P. J. A. Withers, A. P. B. Teles and W. F. B. Herrera (2016). "Legacy Phosphorus and No Tillage Agriculture in Tropical Oxisols of the Brazilian Savanna." **Science of the Total Environment**: 1050-1061.

CHAPTER 2

INFLUENCE OF THE MINERALOGY OF A LITHIC NEOSOL ON MICROBIAL RESPIRATION AND CO2 FLUX

Cassio Ricardo Gonçalves da Costa

Ailson de Lima Marques

Stella Ribeiro Prazeres Suddarth

INTRODUCTION

Litholic neosols

Soils with an A or hystic horizon, resting directly on rock or on a C or Cr horizon or on material with 90% (by volume) or more of its mass made up of rock fragments with a diameter greater than 2mm (gravel, pebbles and boulders), which show a typical lithic or fragmentary contact within 50cm of the soil surface. It admits a B horizon at the beginning of formation, the thickness of which does not fulfil any type of diagnostic B horizon.

Litholic neosols are soils with lithic contact within 50 cm and are normally associated with rock outcrops, with a sequence of horizons A-C-R, A-R, as defined by SiBCS (EMBRAPA, 2006). Because of this, areas where these soils occur have restrictions on human occupation and are generally related to basalts and sandstones.

The morphological characteristics of these soils are practically restricted to those of the A horizon, which varies on average from 0.15 to 0.40 m in thickness, and the colour, texture, structure and consistency depend on the type of material that gave rise to the soil (Silva & Silva, 1997b); in addition, they are quite heterogeneous in terms of physical, chemical and mineralogical attributes (Bognola et al. 2002).

Therefore, considering their characteristics of moderate to high erodibility, wide variation in fertility, small effective depth, presence of impediments to mechanisation and stoniness, these soils require careful management and greater attention from a conservation point of view (Silva & Silva, 1997a). Another aspect that should be taken into consideration is that

7

they are usually sandy, have fragments of rocks and gravel in their body or on their surface and are associated with undulating and strongly undulating terrain, with steep slopes (Silva & Silva, 1985). In addition to their low water storage capacity, these soils also present physical limitations to the growth of the plant root system, thus contributing to further intensifying the effect of water stress on cultivated plants (Cardoso et al. 2002).

According to Sumner (1995), several soils that occur in semi-arid climates have appreciable quantities of weatherable minerals (feldspars, hornblende, plagioclase, calcite and gypsum), which can maintain high activities of calcium, magnesium and sodium ions in solution.

The soil in the area is a Neossolo litolico with a weak A sandy texture, stony and rocky phase, dominated by kaolinite (1:1) and vermiculite and smectite (2:1) hyperxerophilic caatinga gentle undulating relief gneiss substrate. Gneiss generally gives rise to soils with high levels of coarse sand compared to fine sand. Its main mineralogy is more than 20% potassium feldspar, plagioclase and quartz and biotite, with minor constituents such as chlorite, amphibole, garnet, carbonate, epidote and others. The main pre-metamorphic rocks are granite and mica schists.

The area has a granite pegmatite complex, a geology formed by quartz, mica and feldspar and intrusive rocks, which means that in this environment under the semi-arid tropical climate, the mantle of weathering is exposed directly to the surface, In other words, this environment doesn't have time to form soil because the distribution of rainfall is irregular, and there's a torrent of rain, which means that when the minimum amount of soil is formed, the torrent washes away the rock, taking the mantle of weathering with it, which is why neosols are formed.

Soil biological activity - soil respiration

Basal respiration determines microbial activity in the soil, making it possible to quantify the CO_2 released by the metabolic functions of microorganisms, as well as the speed of degradation of a given substrate

(KONRAD & CASTILHOS, 2000). Measures of microbial activity are therefore used as indicators of soil quality (SILVA et al., 2007).

Soil organic matter (SOM) refers to the total organic material in the soil, including identifiable plant residues, animal and microorganism residues, dissolved SOM, root exudates and humic substances. The dynamics of MOS occur through the deposition of organic residues, mainly of plant origin, which contain an average of 40 per cent carbon in their dry matter.

Organic matter as an indicator of soil quality

Most of the thinking about SQ is centred on identifying an index that can serve as an indicator, just as there are indicators for air and water quality. For Islam and Weil (2000), QS cannot be measured directly, but can be inferred from soil properties referred to as soil quality indicator properties (SQIs). The indicators can be distinguished into three large groups: the ephemeral, whose changes occur in a short space of time or are modified by cultivation practices, such as: soil moisture, density, pH, nutrient availability; the permanent, which are inherent to the soil, such as: depth, restrictive layers, texture, mineralogy; and, between these two extremes, are the intermediate indicators, which show a critical influence of the soil's ability to perform its functions, such as: aggregation, microbial biomass, respiration quotient, total and active organic carbon. For these authors, the intermediate indicators are the most important to include in a QS index.

Mechanisms for stabilising organic matter

Physical protection is provided by soil aggregates. It occurs by encapsulating the OM inside the microaggregates or in micropores that are inaccessible to decomposing microorganisms (TISDALL; OADES, 1982; GOLCHIN et al., 1994). The stabilisation time of physically protected organic matter depends directly on the stability of aggregates and their binding agents, which in turn depend on the action of roots and fungal hyphae that are closely related to the presence of plants and/or the addition of residues to the soil

(TISDALL; OADES, 1982). As soil cover is disturbed and removed, the stability and quantity of aggregates decrease, reducing the content of organic matter in the soil. Particulate organic matter (coarse fraction) is mainly protected by physical protection.

Chemical or colloidal interaction comprises intermolecular associations between organic substances and the mineral surface through reactions such as chemical adsorption, electrostatic interactions, hydrogen bridges, ligand exchange and cation bridges. It basically depends on the texture and mineralogy of the soil, with little or no influence from the management system. The greater the specific surface area (SSA) of the soil's mineral particles and the greater the density of functional groups (-O-, -OH, -OH+) on these mineral surfaces, the greater the organo-mineral interaction (CORNEJO; HERMOSÍN, 1996). Clay soils tend to have higher ASE than sandy soils.

Soils with a high concentration of 2:1 expansive minerals (as in Vertisols) or iron and aluminium oxides (as in Latosols) tend to have a higher density of surface functional groups. The organo-mineral interaction can be considered a barrier to microbial attack on the MOS present in the silt and clay fractions, as the enzymes are inactivated by the binding of their functional groups to the mineral surfaces.

Indicators of microbial biomass

Soil microbial biomass data, expressed as carbon and nitrogen content and soil respiration rate, provide indices that make it possible to assess the dynamics of soil organic matter. The activity of microorganisms is generally measured using indicators such as CO_2 released, O_2 absorbed, enzymatic and calorific activities, mineralised N, P, S (DE-POLLI; GUERRA, 2008). The term soil respiration is defined as the absorption of O_2 and/or the release of CO_2 by the living and metabolising organisms in the soil.

A more efficient biomass would be one that loses less C in the form of CO_2 through respiration and incorporates more C into microbial tissues. In samples with the same biomass values, the one with the lowest respiration rate

($<qCO2$) is considered more efficient. Islam and Weil (2000) obtained consistent results stating that as soil quality improves, the metabolic quotient decreases, i.e. the metabolic quotient is negatively related to soil quality and is therefore an indicator of its stress, disturbance or functional imbalance. Basal respiration, which is obtained by measuring the flow of CO_2 evolved from a soil sample, is directly related to soil carbon or microbial biomass.

Microbial biomass and its relationship with soil physical and chemical attributes

Microbial biomass is influenced by the clay content of soils. Clay increases the adsorption of organic compounds and nutrients, provides greater buffering capacity for acidity and protects microorganisms from predators (SMITH; PAUL, 1990). Soils with a high clay content show greater immobilisation of C and N by the microbial biomass (GAMA- RODRIGUES, 1997).

Carbon management index (CMI)

According to (CONCEIÇÃO et al., 2005) the carbon stock of the fraction associated with minerals (CAM) is less sensitive than particulate organic carbon (COP) in detecting changes in organic matter resulting from soil management in the short term. This is because CAM is the most stable fraction of soil OM, consisting of silt and clay size fractions (<53pm), made up of more humified, highly stabilised organic material such as organic compounds left over from the slower degradation process and products of microbial origin

FINAL CONSIDERATIONS

In tropical soils with a high degree of weathering, i.e. a higher concentration of 2:1 clay minerals, oxides and hydroxides, organic matter is reduced and, since soil organic matter is a fundamental component in the functionality of the system, its reduction would influence the physical structure and environmental conditioning for heterotrophic organisms, as well as the soil carbon cycle.

Vermiculite and smectite are expandable 2:1 phyllosilicates that have the capacity to expand and expose the interlayer space, which greatly increases their Specific Surface Area (SSA) and, consequently, their reactivity (Azevedo & Vidal-Torrado, 2009). 2:1 minerals are important because they give soils specific physical and chemical behaviour. The region's soils are dominated by kaolinite (1:1) and vermiculite and smectite (2:1).

Mineral soils with 2:1 clays are soils that have clay materials with particles smaller than 2pm and this strongly affects the dynamics of soil organic matter and consequently the microstructure and pore system of the soil.

Soil mineralogy has a direct effect on the soil's microbial community. The size of the particles, such as 2:1 clay minerals, can influence the pore system, i.e. greater microporosity and thus cause changes in the microbial community, reflected in soil respiration and CO2 release.

BIBLIOGRAPHICAL REFERENCES

Azevedo, A. C. & Vidal-Torrado, P. Smectite, Vermiculite, minerals with hydroxy interlayers and Chlorite. In: Melo, V. de F. & Alleoni, L. R. F. Química e mineralogia do solo. Viçosa, MG: SBCS, 2009.

Barré, P., O. Fernandez-Ugalde, I. Virto, B. Velde and C. Chenu (2014). "Impact of Phyllosilicate Mineralogy on Organic Carbon Stabilisation in Soils: Incomplete Knowledge and Exciting Prospects." Geoderma 235: 382-395.

Bauer, A. and B. D. Velde (2016). Geochemistry at the Earth's Surface, Springer Feng, W., A. F. Plante and J. Six (2013). "Improving Estimates of Maximal Organic Carbon Stabilisation by Fine Soil Particles." Biogeochemistry 112(1-3): 81-93.

Bognola, I. A. et al. Characterisation of the Soils of the Municipality of Carambeí, PR. Rio de Janeiro; Embrapa - Soils, 2002. 75p. - (Embrapa Solos. Boletim de Pesquisa e Desenvolvimento; n. 8).

Cardoso, E. L. et al. Soils of the Urucum Settlement - Corumbá, MS: characterisation, limitations and agricultural suitability. Corumbá: Embrapa Pantanal, 2002. 35p. (Embrapa Pantanal. Documentos, 30).

Conceição, P.C.; Amado, T.J.C.; Mielniczuk, J.;Spagnollo, E. Soil quality in management systems evaluated by organic matter dynamics and related attributes. Revista Brasileira de Ciência do Solo, v.29, p.777-788, 2005.

Cornejo, J.; Hermosín, M. C. Interaction of Humic Substances and Soil Clays. In: PICCOLO, A. (Ed.) Humic substances in terrestrial ecosystems. Amsterdam: Elsevier, 1996.p.595-624.

De-Polli, H.; Guerra, J.G.M. 2008. Soil microbial biomass carbon, nitrogen and phosphorus. p. 263-276. In: Santos, G. de A.; Silva, L.S. da; Canellas, L.P.; Camargo, F.A.O. (Eds.). Fundamentals of soil organic matter: tropical & subtropical ecosystems. 2ª . revised and updated edition. Metrópole, Porto Alegre, RS, Brazil.

EMBRAPA - Brazilian Agricultural Research Corporation. National Soil Research Centre. Brazilian Soil Classification System. 2nd ed. Brasília, 2006. 306p.

Fink, J. R., A. V. Inda, J. Bavaresco, V. Barrón, J. Torrent and C. Bayer (2016). "Adsorption and Desorption of Phosphorus in Subtropical Soils as Affected by Management System and Mineralogy." Soil and Tillage Research 155: 62-68.

Gama-Rodrigues, E.F.; Gama-Rodrigues, A.C. & Barros, N.F. Microbial carbon and nitrogen biomass of soils under different forest covers R. Bras. Ci. Solo, 21:361-365, 1997.

Golchin, A.; Oades, J. M.; Skjemstad, J.O. et all. Study of free and occluded paticulate organic matter in soils by soilid state 13 C CP/MAS NMR spectroscopy and scanning electron microscopy. Australian Journal of Soil Research, Melbourne, v. 32, p. 285-309, 1994.

Hong, H., G. J. Churchman, Y. Gu, K. Yin and C. Wang (2012). "Kaolinite-Smectite Mixed-Layer Clays in the Jiujiang Red Soils and Their Climate Significance." Geoderma 173: 75-83.

Journet, E., Y. Balkanski and S. P. Harrison (2014). "A New Data Set of Soil Mineralogy for Dust-Cycle Modelling." Atmospheric Chemistry and Physics 14(8): 3801-3816.

Konrad, E. E.; Castilhos, D. D. Soil chemical changes and maize growth resulting from the addition of tannery sludge. Revista Brasileira de Ciência do Solo, V.26, n.1, p.257-265,2002.

Islam, K.R. & Weil, R.R. Soil quality indicator properties in mid- atlantic soils as influenced by conservation management. J. Soil Water Conser. , 55:69-78, 2000.

Mikutta, R., C. Mikutta, K. Kalbitz, T. Scheel, K. Kaiser and R. Jahn (2007). "Biodegradation of Forest Floor Organic Matter Bound to
Minerals Via Different Binding Mechanisms." Geochimica et Cosmochimica Acta 71(10): 2569-2590.

Roco, C. A., L. L. Bergaust, L. R. Bakken, J. B. Yavitt and J. P. Shapleigh (2017). "Modularity of Nitrogen-Oxide Reducing Soil Bacteria: Linking Phenotype to Genotype." Environmental microbiology 19(6): 2507-2519.

Rodrigues, M., P. S. Pavinato, P. J. A. Withers, A. P. B. Teles and W. F. B. Herrera (2016). "Legacy Phosphorus and No Tillage Agriculture in Tropical Oxisols of the Brazilian Savanna." Science of the Total Environment 542: 1050-1061.

Silva, J. R. C.; Paiva, J. B. Sediment retention by contour stone ridges on a lithosol slope. Revista Brasileira de Ciência do Solo, Campinas, v.9, p.77- 80, 1985.

Silva, J. R. C.; Silva, F. J. da. Efficiency of contour stone ridges in retaining sediment and improving the properties of a litholic soil. Revista Brasileira de Ciência do Solo, Viçosa, v.21, p.447-456, 1997a.

Silva, J. R. C.; Silva, F. J. da. Productivity of a litholic soil associated with erosion control by contour stone ridges. Revista Brasileira de Ciência do Solo, Viçosa, v.21, p.435-440, 1997b.

Silva, E. E.; Azevedo, P. H. S.; De-Polli, H.; Determination of basal respiration (BSR) and soil metabolic quotient. V. A method for measuring soil biomass. Soil Biology & Biochemistry. 8:209-213, 2007.

Smith, J. L. & E. A. Paul. The significance of soil microbial biomass estimations.

p. 357-396. In J. M. Bollag, & G. Stotzky, (Eds.). Soil Biochemistry. Vol. 6. Marcel Dekker, New York. 367 p.,1990.

Tisdall, J.M. & Oades, L.M. Organic matter and water -stable aggregates in soil. J. Soil Sci., 33:141-163, 1982.

CHAPTER 3

**DEMONSTRATION OF THE METHODOLOGY FOR EXTRACTING
ARBUSCULAR MYCORRHIZAL FUNGI**

Cassio Ricardo Gonçalves da Costa

Ailson de Lima Marques

Stella Ribeiro Prazeres Suddarth

INTRODUCTION

Arbuscular mycorrhizal fungi (AMF) are key microorganisms that make up a large part of the microbial biomass of cultivated soils. Around 80 per cent of plants associate symbiotically with these fungi, which are notably important agents for improving soil quality and crop performance.

A very comprehensive review by AUGÉ (2001) lists 147 scientific papers published up to 1999 that showed benefits provided by arbuscular mycorrhizae in plants subjected to drought or water stress. Although it has been controversial whether or not this benefit is mediated by the increased acquisition of phosphorus or other nutrients, it is already unquestionable, according to the author, that the symbiosis can modify the plant's water balance, at least in some situations and to a certain extent, in a way that is entirely independent of the acquisition of phosphorus. Among the 147 studies listed, 55 studies, through their experimental design, were able to demonstrate that the increase in drought resistance in plants colonised by mycorrhizal fungi occurs independently of the increase in phosphorus absorption capacity, which is also an agronomic benefit provided by the same symbiotic relationship.

Improved soil quality

The extensive development of extraradicular hyphae and the secretion of glomalin by these fungi are important mechanisms for improving the physical, chemical and biological attributes of the soil.

AMF improve soil structure by binding particles together and forming aggregates. This is due to the mechanical effect of the growth of extraradicular hyphae and the secretion of glomalin. This glycoprotein exuded by the fungi

16

has a glue-like property in the soil and there is already a correlation between its content and the stability of aggregates and organic matter.

The growth of hyphae, the exudation of glomalin and the increased production of plant biomass in association with AMF increase the carbon content in the soil, stimulating biological activity as there is more food for the microorganisms. Biologically active soils are suppressive to diseases and nematodes and have a greater presence of processes such as biological nitrogen fixation, phosphate solubilisation, production of plant hormones and mineralisation of nutrients via degradation of waste and organic matter.

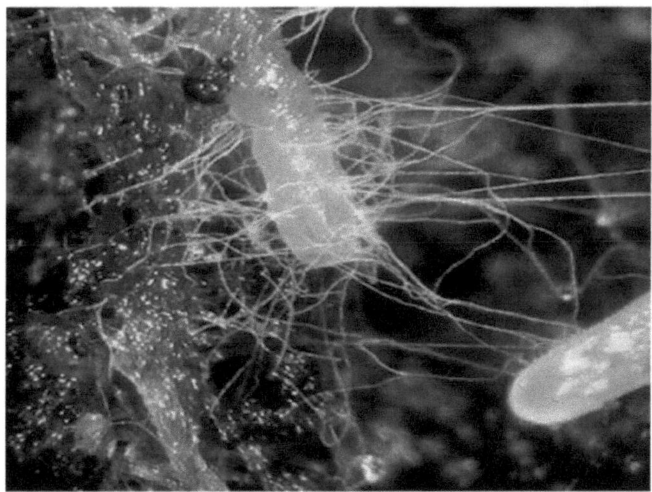

Figure 2- Root colonised by arbuscular mycorrhizal fungi. Source: Denise Mescolotti.

Considering the importance of AMF in maintaining terrestrial ecosystems and the role they play in plant communities, studies on the community structure of these fungi in natural and managed environments are essential. However, comparisons between studies are hampered by differences in sampling effort and the collection method used, and it is important to assess the effect of these factors on knowledge of AMF diversity.

The spores were extracted from the soil using the decantation and wet sieving technique (Gerderman and Nicolson, 1963), followed by centrifugation in water and 50% sucrose (modified from Jenkins, 1964).

METHODOLOGY

Materials needed:

The spores were extracted using the wet sieving method (GERDEMANN; NICOLSON, 1963).

Centrifuge tubes 50 mL;

Soil sieves with 0.42 and 0.053 mm meshes

Petri dishes;

Plastic buckets;

Glass rod;

EQUIPMENT:

Magnifying glass (stereoscopic microscope);

Centrifuge with rotor for 50 mL tubes;

Digital Scale

Blender;

Reagents

Spores can be extracted using a 45, 50 or 60 per cent sucrose solution:

Solution:

Sucrose 45% (450g of refined or crystal sugar for 1 litre of distilled water);

Sucrose 50% (500g of refined or crystal sugar for 1 litre of distilled water);

Sucrose 60% (600g of refined or crystal sugar for 1 litre of distilled water)

1 - Weigh out the caster or crystal sugar;

2 - Add distilled water, stirring or beating in a blender until the sugar is completely dissolved and a transparent yellowish solution is formed, taking care not to exceed a volume of 1 litre;

3 - If you are using crystal sugar, pass the solution through a 0.053 mm mesh sieve to remove impurities from the sugar;

4 - Store the solution in the fridge to prevent fermentation and mould growth

Mycorrhizal fungi extraction procedures:

1 - Wash the sieves with a sponge and detergent

2 - Homogenise the soil samples and separate 50 mL for extraction

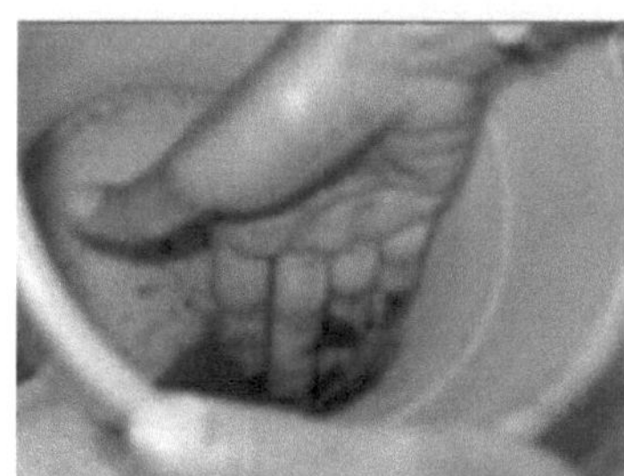

3 - Transfer the 50 mL of sample to a 4 to 5 litre plastic bucket;

4 - Moisten and loosen the soil by pressing it with your fingers against the wall of the bucket, then top up the bucket by suspending the soil with a stick;

5 - Leave to stand for approximately 1 minute so that the coarser soil particles settle and pour the water over the sieves, which should be overlapping (0.42 mm mesh sieve over the 0.053 mm mesh sieve). Only the suspension should be poured off; the coarser soil particles are retained in the bucket.

6 - After the above procedures, the remaining soil in the bucket should be disposed of outside the sink;

7 - Collect the material retained on the 0.42 sieve to check later if there are any spores that have been retained on this sieve;

8 - Using a pissette, transfer the material retained on the 0.053 mm mesh sieve into a 50 ml centrifuge tube.

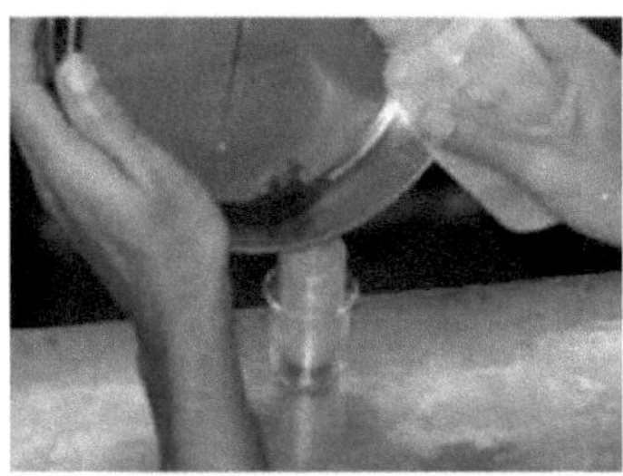

9 - Wash the sieves thoroughly with the sponge and detergent and restart the extraction process with other samples;

10 - Weigh the tubes and standardise them by adding water until they are the same weight;

11 - Centrifuge at 3000 rpm for three minutes;

12 - Remove the tubes and carefully discard the supernatant without moving the solid particles decanted at the bottom of the tube. In this first centrifugation in water, the spores are decanted at the bottom of the tube and the lighter particles remain in the supernatant;

13 - Add the 45, 50 or 60 per cent sucrose solution to the tubes to approximately 40 mL;

OBS: From this point onwards, the extraction process cannot be interrupted and it is recommended that it be carried out as quickly as possible, as the spores may dehydrate due to osmosis;

14 - Suspend the sample in the tube by stirring it with a spatula or glass rod, always taking care to wash the spatula or rod from one sample to the next so that there is no contamination;

15 - Weigh the tubes and standardise them by adding the sucrose solution until they are the same weight;

16 - Centrifuge them once more at 2000 rpm for 2 minutes;

In this procedure, the spores are suspended in the supernatant;

17 - Pour the supernatant carefully over the 0.053 mm sieve so that no decanted particles remain at the bottom of the tube;

18 - Wash the material retained in the sieve with running water to remove the sucrose residue and add it to one side of the sieve, taking care not to throw this material away, as this is where the spores are found;

19 - With the help of a sieve, transfer the material retained in the sieve to a petri dish, Becker or glass to be analysed later under a magnifying glass;

20 - The samples can be stored in glass jars with lids in the fridge, but they should not be left for more than 3 days, as they could germinate or become hyperparasitised;

If the extracted spores are stored in the fridge for another day after removal, they should be washed again in the 0.053 mm sieve to loosen them.

REFERENCES

GERDEMANN, J. W.; NICOLSON, T. H. Spores of mycorrhizal endogone species extracted from soil by wet sieving and decanting. **Transactions of the British Mycological Society**, Cambridge, v. 6, p. 235-246, 1963.

JENKINS, W.R. A rapid centrifugal-flotation technique for separating nematodes from soil. **Plant Disease Report** 48: 692, 1964.

CHAPTER 4

POPULATION SURVEY AND IDENTIFICATION OF ARBUSCULAR MYCORRHIZAL FUNGI IN REPRESENTATIVE SOILS OF PARAÍBA

Cassio Ricardo Gonçalves da Costa

Ailson de Lima Marques

Stella Ribeiro Prazeres Suddarth

INTRODUCTION

Arbuscular mycorrhizal fungi (AMFs) have been a growing focus of research, given the benefits they promote, especially in the soil-plant relationship. These fungi form the most common mutualistic symbiosis in nature, occurring on the roots of most terrestrial plants, promoting improvements in growth, development and increased tolerance to adverse environments (FOLLI-PEREIRA, et al., 2012).

The benefit to plants, mainly due to the increase in phosphorus (P) absorption provided by fungal hyphae (SMITH & READ, 2008), makes the presence of AMFs welcome in the soil. Several studies have been dedicated to surveying and identifying AMFs in different management systems, such as those by Mello et al. (2006), Lima et al. (2007) and Santos and Carrenho (2011).

In the state of Paraíba, the wide geographical variation along its length has made it possible to form various soil classes, ranging from poorly weathered soils to highly weathered soils (Brasil, 1972), thus presenting distinct attributes.

The aim of this study was to survey the population and identify native AMFs from four representative soils in the state of Paraíba: red-yellow Argissolo, yellow Latossolo, LH and NR.

MATERIAL AND METHODS

Soil analyses were carried out at the Soil Organic Matter Laboratory, part of the Department of Soils and Rural Engineering - DSER, at the Centre for Agricultural Sciences - CCA/UFPB, located in the municipality of Areia - PB, at

6°58'12"S and 35°42'15"W, at an altitude of 574.62 m, with an average annual temperature of between 23-24°C and rainfall of 1,400 mm.

Soils from the collection of the Representative Soil Bank of the State of Paraíba (OLIVEIRA, et al., 2006), which can be found in the Department of Soils and Rural Engineering (DSER) at the Federal University of Paraíba (UFPB), were used in this study. This was formed by travelling to the field in areas selected through consultations with Brasil (1972) and Corrêa (2000), looking for diversity in terms of soil classes, at the level of order and suborder of the Brazilian Soil Classification System-SiBCS (Embrapa, 2006), relief, climate and vegetation, as well as representativeness in terms of importance for agricultural production (MONTEIRO, 2008). These soils were collected in 10 municipalities in the state of Paraíba.

For the extraction and quantification of arbuscular mycorrhizal fungi, a 50% sucrose solution was used (500g of refined or crystal sugar to 1L of distilled water); a 0.60 mm mesh sieve over a 0.053 mm mesh sieve, a plastic bucket, the soil was moistened and loosened by pressing it with your fingers against the bucket wall, the volume of the bucket was completed by suspending the soil with the help of a stick, then 50ml of this soil was separated for extraction, this sample was left to stand for approximately 1 minute so that the coarser soil particles could decant, after which the suspended part was poured over the overlapping sieves (0.60 mm mesh sieve over the 0.053 mm mesh sieve).

The material retained on the 0.053 sieve was collected and transferred to 50 ml centrifuge tubes for analysis, the tubes were weighed and water added to bring them to a standardised weight, the tubes were centrifuged at 3000 rpm for 3 minutes, the supernatant was discarded without moving the solid particles decanted at the bottom because the spores are decanted at the bottom of the tube, the sucrose solution was then added to the tubes to make up 40 ml, and they were centrifuged again, this time at 2000 rpm for 2 minutes. The supernatant was carefully poured over the 0.053 mm sieve so that no decanted particles remained at the bottom of the tube; the material retained on the sieve was washed with running water to remove the sucrose residue, and with the aid of a pissette the material retained on the sieve was transferred to a petri dish and analysed under a magnifying glass.

RESULTS AND DISCUSSION

As for the quantification of fungi, the following were found respectively:

1. **Argissolo Vermelho-Amarelo** - 1502 arbuscular mycorrhizal fungi.
2. **Yellow latosol** - 1739 arbuscular mycorrhizal fungi.
3. **Luvissolo Háplico (LH)** - 463 arbuscular mycorrhizal fungi.
4. **Neossolo Regolítico (NR)** - 768 arbuscular mycorrhizal fungi.

As for the identification of the fungi, they were found, respectively:

1. **Red-Yellow Argisol** - (*D. nigra* predominates).

Dentiscutata nigra; Ambispora leptoticha; Dentiscutata erythropa; Acaulospora tuberculata; Gigaspora decipiens.

2. **Yellow Latosol** - (Predominant: *Dentiscutata nigra*) *Dentiscutata nigra; Acaulospora tuberculata; Dentiscutata erythropa; Gigaspora decipiens; Rrizophagus; septoglomus constrictum.*

3. **Luvissolo Háplico (LH)** - (*D. nigra* predominates)

Gigaspora decipiens; Dentiscutata erythropa; Dentiscutata nigra; Glaroideoglomus dunicadum; Acaulospora tuberculata.

4. **Regolithic Neosol (NR)** - (Predominant: *Acaulospora tuberculata*)

Sepdoglomus constrictum; Acaulospora tuberculata; Gigaspora decipiens; Dentiscutata erythropa; Dentiscutata nigra.

The soils used in the analysis showed a wide diversity, as well as a significant number of AMF genera.

The genus *Desntiscutata nigra* is present in all the samples, and predominates in some.

CONCLUSION

The four soil samples analysed showed significant amounts of

Arbuscular Mycorrhizal Fungi, and a variety of genera.

REFERENCES

BRAZIL. Ministry of Agriculture. Research and Experimentation Office. Pedology and Soil Fertility Team. I. Exploratory Soil Survey of the State of Paraíba. II. Interpretation for agricultural use of the soils of the State of Paraíba. Rio de Janeiro: 1972. 683 p. (Boletim Técnico, 15; SUDENE. Série Pedologia, 8).

CORRÊA, M.M. Physical and chemical, mineralogical and micromorphological attributes of soils and agricultural environment of the floodplains of Sousa-PB. Dissertation (Master's Degree) Federal University of Viçosa, Postgraduate Programme in Soils and Plant Nutrition. Viçosa: UFV, 2000. 107p.

CÓRDOBA, A. S., MENDONÇA, M. M., & ARAÚJO, E. F. (2002). Evaluation of the genetic diversity of arbuscular mycorrhizal fungi in three stages of dune stabilisation. **Revista brasileira de ciência do solo**, 26(4), 931-937.

BRAZILIAN AGRICULTURAL RESEARCH COMPANY - EMBRAPA. National Soil Research Centre. Brazilian soil classification system. Rio de Janeiro, 2006. 412p.

FOLLI-PEREIRA, M. S.; MEIRA-HADDAD, LYDICE, S.; BAZZOLLI, D. M. S.; KASUYA, M. C. M. Arbuscular mycorrhiza and plant tolerance to stress. **Revista Brasileira de Ciência do Solo**,36:1663-1679. 2012.

LIMA, R. L. F. A.; SALCEDO, I. H.; FRAGA, V. S. Propagules of arbuscular mycorrhizal fungi in phosphorus-deficient soils under different uses in the semi-arid region of northeastern brazil. **Brazilian Journal of Soil Science.** 31:257-268, 2007.

MELLO, A. H. D.; ANTONIOLLI, Z. I.; KAMINSKI, J.; SOUZA, E. L.; OLIVEIRA, V. L. Arbuscular and ectomycorrhizal fungi in areas of eucalyptus and native grassland in sandy soil. (2006)

MONTEIRO, A. L. MINERALOGICAL CHARACTERISATION OF

REPRESENTATIVE SOILS IN THE STATE OF PARAÍBA. 2010. 116 f. Dissertation (Master's) - Soil and Water Management Course, Centre for Agricultural Sciences, Federal University of Paraíba, Areia, 2010.

OLIVEIRA, F.H.T.; LEAL, J.V.; SANTOS, D.J.; FARIAS, D.R.; ARRUDA, J.A. Banco de Solos Representativos do Estado da Paraíba. In: REUNIÃO BRASILEIRA DE MANEJO E CONSERVAÇÃO DO SOLO E DA ÁGUA, 16, Aracaju, 2006. Proceedings. Aracaju, SBCS, 2006. CD-ROM.

SANTOS, F. E. F.; CARRENHO, R. Diversity of arbuscular mycorrhizal fungi in an impacted forest remnant (Parque Cinquentenário-Maringá, Paraná, Brazil). **Acta Botanica Brasilica**, 25(2), 508-516. 2011.

SMITH, S.E. & READ, D.J. Mycorrhizal symbiosis. 3.ed. London, **Academic Press,** 2008. 785p.

CHAPTER 5

SOIL MANAGEMENT AND VIABILITY OF THE AGROFORESTRY SYSTEM IN THE BANANA HILLS OF BREJO PARAIBANO

Cassio Ricardo Gonçalves da Costa

Ailson de Lima Marques

Stella Ribeiro Prazeres Suddarth

INTRODUCTION

Throughout the process of occupation and land use in the highland marshes, specifically from the 17th to the 20th century in the Paraiba marshes, large areas of forest were replaced by an agrarian mosaic that combines monocultures and polycultures (MARQUES et al., 2014). With a traditionally agrarian economy, the municipality of Areia, the most important in the micro-region, combines management practices that result in soil degradation.

Marques et al. (2014), when zoning the municipality, classified the Santa Maria district and its surroundings as banana hills. This classification refers to the dominance of the hills in the municipality and the replacement of the Atlantic Forest by banana cultivation from the top to the foot of the hills. Areia is the largest banana producer in the micro-region, with production jumping from 3,200 tonnes in 2012 to 9,600 in 2015 on 900 hectares (IBGE, 2015).

The management imposed on soils by agricultural practices and poor management results in their exposure to weathering, leading to the destruction of the pedological and edaphological arrangement (CASSOL & LIMA, 2003); (GARCIA et al., 2005); (GUERRA, 2014), and this has occurred in Brazil due to the prevalence of the agricultural model based on high productivity (RODRIGUES, 2001).

According to Guerra (2014), soil degradation in small municipalities in the interior of Brazil begins with the removal of vegetation, which can be followed by various forms of disorderly occupation, such as: cutting slopes for the construction of houses, roads and railways; agriculture, with the use of burning, various types of mining, excessive irrigation; overgrazing and the use of soil for various types of untreated industrial and domestic waste.

Land degradation, especially in mountainous areas, has been widely discussed in recent years. According to a diagnosis by the Food and Agriculture Organisation of the United Nations (FAO, 2016), in mountainous regions of developing countries, a third of people are at risk of hunger. A source of food and income, agriculture is often one of the main sectors of the economies of these places, which due to their topographical conditions are sensitive to soil degradation factors such as erosion and infertility (dystrophic potential).

In the midst of this, the international body decided to team up with member states and civil society institutions to open up new markets for mountain products. In partnership with the Alliance for the Mountains, the Italian Development Cooperation and the Slow Food organisation, the UN agency has created a label for products from mountain areas that respect the environment and local traditions, using management practices such as agroforestry.

With this in mind, this research aims to characterise the main management practices and the feasibility of incorporating an agroforestry system in a banana-producing mountainous area in the municipality of Areia.

METHODOLOGY

Using data from ongoing university extension research, an assessment was made of the levels of management employed in the area, based on Ramalho Filho and Beek (1997), through the application of pre-structured questionnaires, according to the methodology of (MUGGLER, 2006); (FRANCO, 2005), with 65 farmers from the family farming system in mountainous areas.

The questionnaires were processed using the Statistical Method of Simple Random Sampling (LITTLE, 1987), with the aim of obtaining a variety of information, including: intellectual level; income; technological level used in crops; identification of soil fertility problems; recognition of Permanent Preservation Areas (PPAs), especially hilltops and riparian vegetation, and the feasibility of incorporating the agroforestry system into production areas.

RESULTS AND DISCUSSION

According to the material analysed in the content analysis, 70% of the interviewees adopt system A (primitive) and 30% B (intermediate), where a basic level of technology prevails, with little capital investment for the management, improvement and conservation of the conditions of the land and crops. Farming practices depend on manual labour, and some animal traction may be used with simple agricultural implements.

Level (A) is called primitive, since it brings together agricultural practices that reflect a low technical and cultural level. There is practically no application of capital to the management, improvement and conservation of land and crop conditions. Level (B) is called intermediate, as it brings together farming practices that reflect a medium technological level. There is a modest application of capital and research results for the management, improvement and conservation of land and crop conditions.

When asked about the APPs (5%), they preserve riparian vegetation on at least 5 metres of marginal land, 2% preserve hilltops and 1% preserve riparian vegetation and hilltops together. When asked about incorporating the agroforestry system, none of the interviewees knew anything about it and 20% were interested in incorporating the system. All the interviewees recognise problems with erosion.

The slope of the area varies between gently undulating surfaces (4-8%) in the foothills and strongly undulating (>30%) to steep (>75%) on the tops. The steeply sloping areas corroborate the problem of water erosion, which is further accentuated by rainfall levels (1200-1300mm/8-9 months of orographic rain) and poor management, which intensifies erosion. According to EMBRAPA (2010), areas with a slope of more than 30 per cent are considered unsuitable, as strict soil erosion control measures are required.

According to Pimentel et al. (1995), erosion is considered one of the greatest threats to sustainable development and the productive capacity of agriculture, due to the transport of nutrients, loss of organic matter and pesticides, causing a sharp decline in the energy exchange capacity of soils and plants. It is estimated that around 1.5 billion hectares (10% of the earth's surface) have already been irreversibly degraded by erosion processes.

CONCLUSIONS

In the face of environmental degradation, which includes deforestation of hilltops and riparian vegetation, poor management, exhaustive use of the soil and erosion and fertility problems due to the influence of the mountainous to sloping terrain, the incorporation of the agroforestry system is of fundamental importance for this area.

By incorporating such a system, in addition to revegetating hilltops and riverbanks, slope terracing, a native vegetation-banana consortium and top-to-top ecological corridors could be incorporated.

REFERENCES

CASSOL, E. A.; LIMA, V. S. Erosion in gullies under different types of soil preparation and management. Pesquisa Agropecuária Brasileira, v. 38, n. 1, p. 117-124, 2003.

GARCIA, G. J.; ANTONELLO, S. L.; MAGALHÃES, M. G. M. New version of the land evaluation system - SIAT. Engenharia Agrícola, v. 25, n. 2, p. 516-529, 2005.

GUERRA, A.J.T.. Soil Degradation - Concepts and Themes. In: Antonio Jose Teixeira Guerra;Maria do Carmo Oliveira
Soil Degradation in Brazil. 1ed.Rio de Janeiro: Bertrand Brasil, 2014, v. 1, p. 15-50.

LITTLE, R. J. A.; RUBIN, D. B. Statistical Analysis with Missing Data, Wiley, New York, 1987.

MARQUES, A.L.; SILVA, J.B; SILVA, D.G. HUMID REFUGES IN THE

SEMI-ARID REGION: A STUDY OF THE HIGHLAND MARSHLAND OF AREIA-PB. Geotemas Magazine. V.4, n.2. P.17-31, 2014.

MUGGLER, C.C.; SOBRINHO, F.A.P.; MACHADO, V.A. Educação em solos: princípios, teoria e métodos. Revista Brasileira de Ciência do Solo, v.30, p.733-740, 2006.

PIMENTEL, D. Soil erosion. Environment, v. 39, n. 10, p. 45, 1997

RAMALHO-FILHO, A.; BEEK, K. J. Sistema de avaliação da aptidão agrícola das terras. 3. ed. Rio de Janeiro: EMBRAPA- CNPS, 1995. 65 p.

RODRIGUES, R. Agricultura e agronomia. Estudos Avançados, v. 15, n. 43, p. 289-302, 2001.

CHAPTER 6

LAND USE AND CHARACTERISATION OF RURAL AREAS IN THE BUFFER ZONE OF MATA DO PAU FERRO STATE PARK, AREIA-PB

Ailson de Lima Marques

Cassio Ricardo Gonçalves da Costa

Stella Ribeiro Prazeres Suddarth

INTRODUCTION

The current concern about anthropogenic actions on the biosphere has led man to rethink nature's resilience. As a result, mankind has realised the need to plan environmental actions and redefine territorial preservation policies and resource extraction limits (FLORIANO, 2004).

It was from the 1970s and 1980s onwards that the concepts of territorial planning brought together environmental issues, reflecting in environmental planning actions, defined as a process of justice for actions geared towards potential, local culture and the carrying capacity of nature, seeking to maintain the quality of the physical, biological and social environment (SANTOS, 2003).

Conservation units (CUs) have been incorporated as part of these environmental planning actions, as they are areas in which certain land use restriction measures are applied in order to protect a certain local natural or cultural feature. They emerged with different peoples: the Assyrians, a Middle Eastern civilisation, established reserves even before the birth of Christ; in the Middle Ages, Europe had numerous nature protection areas under the responsibility of the king; and in India, protected areas have existed for over a century (MACHADO, 2014).

Two concepts are provided for in the Brazilian Forest Code Law No. (12.651/12) and in the National System of Conservation Units (SNUC) Law No. (9985/2000), which have been widely discussed at scientific congresses, in the Chamber of Deputies and in the Federal Senate, having repercussions on the discussion of Conservation Units, through licensing and environmental monitoring.

The first is Buffer Zones (BZs) and the second is Ecological Corridors (ECs), which represent an effort to secure and conserve biodiversity beyond the boundaries of the PAs.

The ZAs are peripheral delimitations that go beyond the political boundaries of the UCs, they are not part of the unit, but they promote ecological stability that is safe for its biodiversity. ECs, on the other hand, seek to mitigate the effects of the fragmentation and insularisation of natural environments, which harm or even make it impossible to maintain biodiversity due to the reduction and isolation of ecosystems, through the connectivity of preserved areas.

Among the key authors in this discussion are: Bierregaard, et al. (1992), Campos & Agostinho (1997); Primack & Rodrigues (2001), Ribeiro et al., (2010) and Machado (2014).

The legality of the buffer zone

The establishment of the ZA (monitoring perimeter) is contained in Brazilian legislation and is defined in Article 2, item XVIII of the Law (No. 9.985/2000) as: "the surroundings of a conservation unit, where human activities are subject to specific rules and restrictions, with the aim of minimising negative impacts on the unit".

The establishment of this surrounding area is also contained in CONAMA Resolution 13/90, which states that all anthropogenic activities that could affect the biota within an estimated radius of up to ten kilometres (10 km) from the areas surrounding the conservation unit must be licensed and monitored by the environmental agency administering the conservation unit. The radius that forms the ZA is accommodated within the boundaries of the PAs, taking into account geographical, biological and social parameters, which may be contained in the Ecological-Economic Zoning of the area, if any.

Forman & Godron (1986) and Ribeiro et al. (2010) argue that the edges of protected areas are sensitive to degrading actions that directly and indirectly influence the protected area. This anthropogenic action on the edges generates the so-called Edge Effects (EB), which are changes in the diversity of the flora, which tends to lead to a loss of faunal diversity in the areas adjacent to the PA.

They are also divided into physical effects and direct and indirect biological effects. Physical effects involve changes in environmental climatic factors, such as the formation of a warmer microclimate. Direct biological effects involve changes resulting from physical effects, such as the spread of species adapted to increased solar radiation. Indirect effects involve changes in the ecological interactions between species, such as predation, parasitism, herbivory, competition, seed dispersal and pollination (RODRIGUES, 1993).

Based on this discussion, this research aims to: map land use and occupation, simulate and establish the buffer zone and characterise rural spaces.

MATERIAL AND METHODS

Study area

This research was carried out in the Mata do Pau-Ferro State Park, located in the municipality of Areia (Figure 1). The municipality is located in the Agreste mesoregion and Brejo microregion of Paraíba, in north-eastern Brazil.

Figure 1- Location of the Mata do Pau-Ferro State Park, Areia-PB.

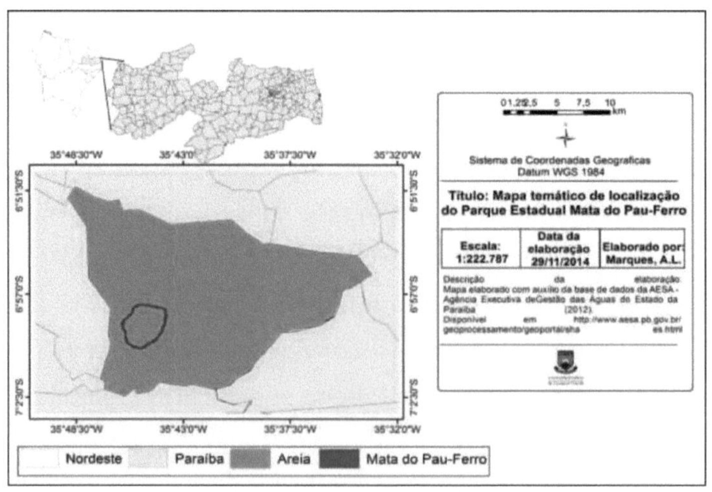

It has an area of 269.4 Km^2 , an estimated population of 23,829,00 inhabitants and a population density of 88.42 inhabitants per square kilometre (IBGE, 2015).

The park is located in the south-western part of the state and covers 600 ha, under a hot and humid tropical "As" climate (Koppen). The geology of the area is dominated by the morphostructural unit of the interplanaltic depression of the Borborema Plateau in Paraiba, with capping of the Serra dos Martins Formation.

The relief is of flat-topped hills, valleys, floodplains and dissected areas to windward. The soils vary from Latossolos, Argissolos and Nitossolos to Neossolos and Luvissolos. The hypsometry varies from 164 to 635m within the domain of the vaca brava sub-basin, with a predominance of high-altitude and exposed swamp forests from the Atlantic Forest biogeographic domain (Marques et al., 2014).

It is the most biodiverse highland wetland park in the country, where 309 species of Angiosperms were listed, distributed in 84 families: Rubiaceae (24), Malvaceae (21) Asteraceae (14), Convolvulaceae (12), Solanaceae (16) and Fabaceae (12), while the genera were Sida (11), Solanum (10), Ipomoea (5), Piper (6), Psychotria (5), Vernonia (5) and Senna (6).

Technical procedures

Use and occupation and buffer zone

The Landsat 8 programme image from 16/08/2014, acquired from the USGS-US Geological Survey platform, was processed. It underwent Radiometric and reflectance calibration in the ArcGIS 10.5 software. Subsequently, the Maximum Likelihood (MAXVER) method of ArcGIS 10.5 was applied, and spectral signatures were created to obtain land use and occupation classes: Crops, Pastures, Forest Vegetation and River Waters.

The UC's territorial limits were then mapped using shapefile vectors, and a simulation of one (1) kilometre of marginal stratification was established using the ArcGIS 10.5 buffer method to delimit the ZA.

Characterisation of rural areas

The characterisation of rural areas was an adaptation, for the municipality and the buffer zone, of the new IBGE methodology (2017), which uses cross-referenced georeferenced data in the form of shapefiles of demographic density, population and rural dwellings, rural activities and production, the Agricultural Census and the Rural Environmental Registry.

With the aid of geoprocessing, different spectral environments of use and occupation were formed and reclassified into rural spaces: diversified crops, monoculture crops, pastures, bare soil, fallow areas and water bodies. Photographic cuttings were then made and the reclassification of rural spaces was validated.

The ArcGIS 10.5 software is licensed to the Multiuser Laboratory of Information Technologies Applied to the Human Sciences (LabINFO), Digital Cartography, Geoprocessing and Remote Sensing sector (CADIGEOS), of the Postgraduate Courses at the Humanities Centre of the Federal University of Campina Grande.

RESULTS AND DISCUSSIONS

Use and occupation and buffer zone

By mapping use and occupation, it was identified that the Park has 600 ha, of which 578 have a spectral response of climax forest vegetation and 22 represent clearings with vegetation in various successional stages and pastures (Figure 2). This situation corroborates Barbosa et al. (2004).

According to the authors, this park suffers from anthropogenic pressure, with subsistence agricultural crops such as beans, maize and manioc being planted in its interior, which are cyclically abandoned, forming clearings with scrubland of different successional stages, some of which expose the bare soil to sunlight, making regeneration difficult.

The buffer zone comprises 560 ha and there are no ecological corridors. There are areas of forest vegetation totalling 160.5 ha, as well as 108.8 ha of pasture and 288 ha of various crops (subsistence and monoculture). The buffer zone is home to farms, smallholdings and the districts of Muquem and Cepilho, as well as small communities such as Chã de Jardim, Vaca Brava and Sítio Macacos, which have a population of around 3,000 people.

In the buffer zone, all economic and social activities should be monitored and preservationist environmental education built, but these surrounding communities have been responsible for deforesting the remnants that border the park and enter the unit for deforestation, the implementation of various crops and pastures.

This situation was also described by Silva et al. (2006), who found the presence of anthropogenic activities in this environment, such as housing, disposal of inorganic waste, deforestation, capture of wild animals and cultivation and grazing areas.

According to the authors, the practice of removing plants from the reserve for commercial purposes, such as trees, shrubs, orchids and bromeliads, was also observed.

Figure 2- Land use and occupation and simulation of the buffer zone of the Mata do Pau-Ferro State Park, Areia-PB.

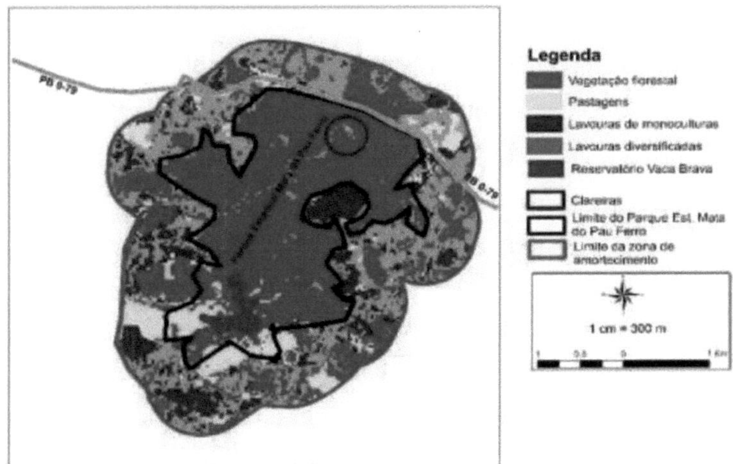

Rural areas

The rural spaces identified (Table 1) are a representation of the spatial distinctions in land use, tending to be linked to the types and values of

37

investments made in the countryside.

Table 1- Classification of rural areas in the buffer zone of the Mata do Pau-Ferro State Park, Areia-PB.

Rural areas within the ZA	Nomenclature	Activities	ha	%
Ploughing	Miscellaneous (Subsistence)	Planting beans, maize, manioc, vegetables, coconuts, bananas, etc.	190	34
	Monoculture regime	Planting sugar cane, flowers and roses for rural tourism purposes	98	17
Pastures	Livestock purposes	Brachiaria and elephant grass	108,8	19
Forest vegetation	Climax forest	Remnants of seasonal forest under pressure	160,5	28
River water	Damming	X	X	X

Corroborating this, when zoning the municipality of Areia, in the area under study, Marques et al. (2014) identified two zones of anthropisation and classified them as: Mata do Pau-Ferro, characterised by open Ombrophilous forest, an area of intense disturbance; and Mares de Morros, characterised by highly disturbed Semideciduous Forest. semideciduous forest.

In view of these sample cut-outs, there is a need for land uses to be licensed and monitored, and so discussions have arisen about the importance of the ZA in PAs.

CONCLUSIONS

No environmental monitoring or licensing mechanisms were identified in the Park or in the ZA simulation. In summary, in terms of land use, the Park has an environment of insularisation in terms of land use and conditions.

Throughout the ZA there is fragmentation with remnants of forests under pressure and in various successional stages. These areas are commonly cleared for agricultural use (fallow and crop rotation), thus creating a socio-environmental conflict. Monitoring and licensing in these circumstances would allow the Permanent Preservation Areas to be restored and the most sensitive areas to be re-vegetated with native species, while also encouraging the return of pollinating and dispersing species, which would help with the environmental systemic balance of the entire ZA.

REFERENCES

ALLEN, R.; TASUMI, M.; TREZZA, R. **SEBAL (Surface Energy Balance Algorithms for Land).** Advanced Training and Users Manual - Idaho Implementation, version 1.0, 2002.

BIERREGAARD, R. O.; Jr., T. E. L.; Kapos, V.A. A. S.; Hutchings, R. W. The biological dynamics of tropical rainforest fragments. **BioSciences**: 42:859-866, 1992

CAMPOS, J.B.; AGOSTINHO, A.A. Paraná River biodiversity flow corridor: a proposal for integration and environmental protection of threatened ecosystems. In: Brazilian Congress of Conservation Units, 1. 1997. Curitiba, PR, **Proceedings**... Curitiba, p.645-657.

FORMAN, R.T.T. & GODRON, M. **Landscape ecology**. John Wiley & Sons, New York. 1986. 620 p.

FLORIANO, E. P. **Environmental planning**. Santa Rosa: ANORGS, 2004. 58 p.

IBGE - Brazilian Institute of Geography and Statistics. IBGE Cities Portal, 2015. **Areia-PB**. Available at: http://www.cidades.ibge.gov.br/painel/painel.php?lang=&codmun=2 50110&search=paraiba%7Careia%7Cinfograficos:-dados-gerais-do-municipio. Accessed on: 21 January 2015.

MACHADO, C. C. C. **Surface changes in Catimbau National Park (PE - BRAZIL): biophysical indicators and human influence**. PhD thesis submitted to the Postgraduate Programme in Geography at the Federal University of Pernambuco. 2014, 194p.

MARQUES, A.L.; SILVA, J.B; SILVA, D.G. Humid refuges of the semi-arid: a study on the altitude and exposure marsh of Areia-PB. **Geotemas Magazine**. V.4, n.2. P.17-31, 2014.

MARKHAM, B. L. & BARKER, J. B. Thematic mapper band pass solar exoatmospherical irradiances. **International Journal of Remote Sensing**. v.8, n.3, p.517-523, 1987.

PRIMACK, R.B. & E. RODRIGUES. **Conservation Biology**. Londrina, E. Rodrigues, 2001, 328p.

RIBEIRO, M. F.; FREITAS, M. A. V.; COSTA, V. C. O desafio da gestão ambiental de zonas de amortecimento de unidades de conservação. In: LATIN AMERICAN SEMINAR ON PHYSICAL GEOGRAPHY, 6, 2010, Coimbra. **Proceedings**...VI SLAGF. Coimbra, Portugal: [s.n.].

RODRIGUES, E. **Ecology of forest fragments along a gradient of urbanisation in Londrina-PR**. 1993.
110 f. Dissertation (Master's in Ecology) - Federal University of São Carlos, São Carlos, 1993.

SANTOS, R. F. **Princípios de planejamento ambiental**. Campinas, SP: Oficina de Textos, 2003. 247 p.

SCHOWENGERDT, R.A. **Techniques for image processing and classification in remote sensing**. London: Academic Press, 1980.

CHAPTER 7

ENVIRONMENTAL AWARENESS OF PERMANENT PRESERVATION AREAS IN THE MUNICIPALITY OF AREIA-PB: THE CAR UNDER DISCUSSION

Cassio Ricardo Gonçalves da Costa

Ailson de Lima Marques

Stella Ribeiro Prazeres Suddarth

INTRODUCTION

Reflection on the relationship between man and nature in this century is intrinsically marked by the degradation of the environment and associated ecosystems. Environmental education has thus emerged as an important element of society and economic relations. This dimension of sustainability and preservation has emerged mainly in the basic education and university sectors, as well as in the agricultural and business sectors (TAUCHEN and BRANDLI, 2006).

In this discussion, one of the most important weapons in the Brazilian Forest Code (Law 12.651/12) is the Permanent Preservation Areas (APPs), which are priority areas of effective protection, whether or not they are covered by native vegetation, with the environmental function of preserving water resources, the landscape, geological stability and biodiversity, facilitating the gene flow of fauna and flora, protecting the soil and ensuring the well-being of human populations. Thus, assessing environmental awareness is one of the most important ways of knowing the profile and degree of awareness of social groups and organisations on issues involving preservation and sustainability (DIAS, 2003).

With this in mind, the New Forest Code was created with the main aim of resolving the environmental liabilities of rural landowners, i.e. regularising properties that were in breach of the law. To this end, programmes and tools were created to better articulate environmental regularisation procedures, with the CAR being the mechanism most expected to be effective in managing and monitoring APPs (LAUDARES et al., 2014).

This work seeks to investigate and characterise the environmental awareness of small and medium-sized farmers regarding the preservation of APPs on their rural properties in the municipality of Areia.

METHODOLOGY

Using data from ongoing university extension research, an assessment was made of the profile of environmental awareness, based on Dias (2003) and UNESCO (2005), applied to small and medium-sized farmers using pre-structured questionnaires, according to the methodology of MUGGLER (2006) and FRANCO (2005), with 40 farmers standardised in the Rural Environmental Registry (CAR).

The questionnaires were processed using the Simple Random Sampling Statistical Method (LITTLE, 1987) in order to obtain a variety of information, including: intellectual level; income; types of crops; and recognition of Permanent Preservation Areas (PPAs), especially hilltops and riparian vegetation.

RESULTS AND DISCUSSION

In the municipality, due to the geographical characteristics of high altitude marshes, such as mountainous to rugged terrain with dissected hills and steep slopes under a humid climate, the main forms of APPs are hilltops and riparian vegetation (watercourses).

In this context, when asked what APPs are, 30% were able to tick the correct box, 40% ticked incorrectly and 30% chose not to tick. When asked about maintaining PPAs and Legal Reserves (LRs) in the long term, 80% said that they could interfere with the area.

Taking into account that all the properties have the CAR and that the majority of their owners are unaware that it is mandatory to maintain the APPs, the preservation of the CAR areas becomes ineffective. This finding is corroborated by the fact that 100% of them recognised that the state's environmental agency, which is responsible for overseeing legal compliance with the CAR, does not have the appropriate physical structure.

Also in this discussion, according to Sparoveck et al. (2011), the new Brazilian Forest Code poses the risk that the government will not have effective means to control the accounting of APPs, as well as supporting the calculation of APPs in the calculation of RLs, leaving the area to be preserved variable and which may not correspond to a real APP.

According to Laudares et al. (2014), because the CAR is in the process of being improved and implemented, it has not been instituted throughout Brazil. As such, there is still legal uncertainty as to whether or not the Legal Reserve must be registered.

CONCLUSIONS

In the municipality, the farmers consulted have built up a negative profile in terms of awareness of APPs, which may be imbricated by the legal opening given by RLs to protect fauna and flora in a chosen location, often without biological technical criteria, rather than a real APP.

A sense of impunity predominates among the public consulted, since they don't believe in the role of the state as a supervisory and regulatory body.

These findings corroborate the entire scientific community that is currently discussing the subject.

REFERENCES

DIAS, G.F. Educação Ambiental: princípios e práticas. 8.ed. São Paulo: Gaia, 2003.

JACOBI, P. Environmental education, citizenship and sustainability. Cadernos de Pesquisa, n. 118, p. 189-205, 2003.

LAUDARES, S. S. A. ; SILVA, K. G. ; BORGES, L. A. C. . Rural Environmental Registry: an analysis of the new tool for environmental regularisation in Brazil. DEVELOPMENT AND THE ENVIRONMENT (UFPR) , v. 31, p. 111-122, 2014.

LITTLE, R. J. A.; RUBIN, D. B. Statistical Analysis with Missing Data, Wiley, New York, 1987.

MUGGLER, C.C.; SOBRINHO, F.A.P.; MACHADO, V.A. Educação em solos: princípios, teoria e métodos. Revista Brasileira de Ciência do Solo, v.30, p.733-740, 2006.

RODRIGUES, R. Agricultura e agronomia. Estudos Avançados, v. 15, n. 43, p. 289-302, 2001.

SPAROVEK, G.; BARRETTO, A.; KLUG, I.; PAPP, L.; LINO, J. A revisão do Código Florestal brasileiro. Novos Estudos- CEBRAP, São Paulo, 89, 111-135, 2011.

TAUCHEN, J.; BRANDLI, L.L. Environmental management in Higher Education Institutions: a model for implementation on a University Campus. Gestão & Produção, v. 13, n. 3, p. 503-515, 2006.

Buy your books fast and straightforward online - at one of world's fastest growing online book stores! Environmentally sound due to Print-on-Demand technologies.

Buy your books online at
www.morebooks.shop

Kaufen Sie Ihre Bücher schnell und unkompliziert online – auf einer der am schnellsten wachsenden Buchhandelsplattformen weltweit! Dank Print-On-Demand umwelt- und ressourcenschonend produziert.

Bücher schneller online kaufen
www.morebooks.shop

Printed by Books on Demand GmbH, Norderstedt / Germany